FISH

BY ERIC GERON

Children's Press
An imprint of Scholastic Inc.

A special thank-you to the team at the Cincinnati Zoo & Botanical Garden for their expert consultation.

Copyright © 2023 by Scholastic Inc.

All rights reserved. Published by Children's Press, an imprint of Scholastic Inc., *Publishers since 1920.* SCHOLASTIC, CHILDREN'S PRESS, and associated logos are trademarks and/or registered trademarks of Scholastic Inc.

The publisher does not have any control over and does not assume any responsibility for author or third-party websites or their content.

No part of this publication may be reproduced, stored in a retrieval system, or transmitted in any form or by any means, electronic, mechanical, photocopying, recording, or otherwise, without written permission of the publisher. For information regarding permission, write to Scholastic Inc., Attention: Permissions Department, 557 Broadway, New York, NY 10012.

Library of Congress Cataloging-in-Publication Data Available
Identifiers: LCCN 2022001738 (print)
ISBN 9781338836554 (library binding) | ISBN 9781338836561 (paperback)

10 9 8 7 6 5 4 3 2   23 24 25 26 27

Printed in China   62
First edition, 2023

Book design by Kay Petronio

Photos ©: cover top, back cover left, 1 top, 2 right: Doug Perrine/Blue Planet Archive; cover bottom, back cover right, 1 bottom, 2 left: D. R. Schrichte/Blue Planet Archive; 4 top: Marty Wakat/Dreamstime; 4 bottom left: lindsay_imagery/Getty Images; 4 bottom right: Norbert Wu/Minden Pictures; 5 top: Wafue/Getty Images; 5 center: Mark Newman/Getty Images; 5 bottom: Giordano Cipriani/Getty Images; 6: Steven Kovacs/Blue Planet Archive; 7: D. R. Schrichte/Blue Planet Archive; 8-9: D. R. Schrichte/Blue Planet Archive; 8 inset: Gregory G. Dimijian/Science Source; 10-11: Marko Steffensen/Blue Planet Archive; 24-25: Chris Newbert/Minden Pictures; 26: Kyle Haasbroek/Getty Images; 27: Pete Oxford/Minden Pictures; 28-29 all: Pete Oxford/Minden Pictures; 30 top left: Doug Perrine/Blue Planet Archive; 30 top right: D. R. Schrichte/Blue Planet Archive; 30 bottom center: Rene Gazzola/EyeEm/Getty Images; 30 bottom right: Alexis Rosenfeld/Getty Images.
All other photos © Shutterstock.

DWARF SEAHORSE

SAILFISH

CONTENTS

Meet the Fish	4
#10 Slowest Fish: Dwarf Seahorse	6
Dwarf Seahorse Close-Up	8
#9: Anglerfish	10
#8: Flounder	12
#7: Sunfish	14
#6: Moray Eel	16
#5: Whale Shark	18
#4: Manta Ray	20
#3: Great White Shark	22
#2: Barracuda	24
#1 Fastest Fish: Sailfish	26
Sailfish Close-Up	28
Fish Fast and Slow	30
Glossary	31
Index	32

MEET the FISH

Welcome to the underwater world of fish! This group of animals includes eels, seahorses, sharks, stingrays, and tropical fish. Fish have a lot in common. All fish live in water. They swim with fins. They breathe through **gills**. Most fish have **scales** covering their bodies. Fish have **backbones**. They are **cold-blooded**. This means their body temperatures change with their surroundings.

How Fish Move

All fish can move! But... how do they do it? Some fish have a pair of **pectoral fins**, one on either side of their bodies. These fins help them steer through the water with ease. The **dorsal fin** helps them keep their balance. Fish move in many ways! Get ready to discover how 10 fish can travel, from the slowest to the fastest!

FACT Scientists who study fish are called **ichthyologists** (ik-thee-AH-luh-jists).

#10 Slowest Fish: DWARF SEAHORSE

The dwarf seahorse is the world's slowest fish. This slowpoke of the sea moves at about 0.01 miles per hour (0.02 kph). To compare, the average person can swim at 2 miles per hour (3.2 kph).

Why so slow? Dwarf seahorses are not shaped like many other fish. Their bodies are upright, and their faces look like a miniature horse. The dwarf seahorse is a poor swimmer. It has a small fin in the middle of its back and small fins on either side of its head. The dwarf seahorse tends to stay in one place to survive. It latches on to seaweed or other plants so it can eat and not get swept away.

FACT Seahorses can change colors.

DWARF SEAHORSE CLOSE-UP

A baby seahorse is called a **fry**. Each fry is only about the height of two pennies stacked together!

BONY PLATES
Most seahorses don't have scales. Instead, they have hard, bony plates.

FINS
Seahorses have fins on either side of their head and one longer fin on their back.

TAIL
The tail is perfect for wrapping around plants to keep the seahorse in place.

FACT: A group of seahorses is called a **herd**.

EYES
The eyes can move in two different directions to help look out for **predators** and prey.

SNOUT
Seahorses don't have teeth! They slurp up food, like small **crustaceans**, through their long, tubelike snouts.

NECK
A dwarf seahorse can move its neck quickly to snatch up prey in its mouth.

POUCH
The male seahorse, not the female seahorse, has a pouch to carry its babies.

#9 ANGLERFISH

The anglerfish lives in the deep sea. It moves slowly through the water because its body is so wide. The anglerfish can reach a top speed of only 0.24 miles per hour (0.39 kph). It swims in the dark. Females have a special glowing fin that attracts prey to its welcoming light. They wave it back and forth like a fishing rod, waiting to chomp down on fish that swim toward it.

FACT: The anglerfish can swallow prey twice its size.

FACT There are about 200 types of anglerfish.

Both eyes are on the same side of the flounder's face! FACT

#8 FLOUNDER

Sometimes, a flounder can be nearly impossible to spot! Why? This flat fish can bury itself into the sand on the seafloor with only its eyes sticking out. The flounder can also change colors to **camouflage** itself, or blend into its surroundings. When it's time to swim, it uses its body and tail fin to move through the water. Flounder can travel at only 0.36–0.76 miles per hour (0.58–1.2 kph).

#7 SUNFISH

From the looks of its pointed dorsal fin slicing through the water, some may mistake a sunfish for a shark! The sunfish's back fin doesn't grow with its body though, which makes this fish a slow and wobbly swimmer. It can move at only 2 miles per hour (3.2 kph). Sometimes, when the sunfish jumps, or **breaches** the surface, it can reach a height of up to 10 feet (3 m). Talk about getting air!

FACT At a whopping 5,000 pounds (2,268 kg), the sunfish is the biggest bony fish in the sea.

FACT While moray eels have poor eyesight, they have an excellent sense of smell.

#6 MORAY EEL

The moray eel may look and move like a snake as it swims through the water, but it is a type of fish. Its fins are joined together to look like one long dorsal fin. It can be up to 8 feet (2.4 m) long and weigh more than 65 pounds (29.4 kg). The moray eel has sharp teeth. It waits for prey to swim by, then it strikes. It can move up to 2.4 miles per hour (3.9 kph).

FACT: Moray eels live in shelters made of rocks or coral.

#5 WHALE SHARK

The whale shark is the largest fish in the world. It can grow up to 40 feet (12.2 m) long and weigh as much as 41,200 pounds (18,688 kg). That's as big as a school bus! Even though it is large and powerful, it is also a slow-moving fish. A whale shark can swim up to only 3 miles per hour (4.8 kph).

FACT Each whale shark is covered in a unique pattern of white spots.

Manta rays may live alone or in groups. **FACT**

#4 MANTA RAY

FACT: Manta rays can live for up to 50 years!

The average manta ray can reach speeds up to 22 miles per hour (35.4 kph). To compare, a fast human can run 10-15 miles per hour (16.1-24.1 kph). Manta rays look like large cloaks cruising through the water. They flap their pectoral fins up and down to allow them to move. Their largest **wingspan** is around 30 feet (9.1 m) from fin tip to fin tip. It allows these graceful giants to glide through the water.

#3 GREAT WHITE SHARK

FACT: A baby great white shark is called a pup.

The great white shark is a powerful predator. It can weigh up to 5,000 pounds (2,268 kg) and grow up to 20 feet (6.1 m) long. This fish usually swims slowly to save energy. But when the great white shark chases prey, it can swim up to 35 miles per hour (56.3 kph). This shark is built for speed. Its strong tail propels it forward, and its dorsal fin cuts through water with ease. It can even breach to catch prey, like seals.

FACT Great white sharks are found in oceans all around the world.

23

FACT While most barracuda are around 2 feet (0.61 m) long, the largest found was 7 feet (2.1 m) long.

#2 BARRACUDA

A barracuda is a **carnivore** that hunts at night. It eats other fish. It chases prey. Having a long, thin body lets the barracuda rip through the water like an arrow. It has been recorded swimming at 36 miles per hour (57.9 kph).

#1 Fastest Fish: SAILFISH

The sailfish is the fastest fish ever recorded. This sleek-looking fish can reach top speeds of up to 68 miles per hour (109.4 kph). That is about the same speed as the world's fastest sailboat!

When it hunts food, the sailfish zips through the water like a torpedo! To reach its record-breaking speeds, a sailfish folds back its fins. This makes its shape more streamlined so that it can move even faster. More streamlined means greater speed.

FACT Sailfish can live as long as 15 years.

DORSAL FIN

The sailfish has a gigantic dorsal fin on its back. This fin helps it round up shoals, or large groups of fish that swim together, while hunting.

BODY

At an average size of 10 feet (3 m) and 120–220 pounds (54.4–99.8 kg), the sailfish has a sleek and slender body built for slicing through water.

SCALES

These scales can change color when the sailfish gets excited or wants to confuse its prey.

SAILFISH CLOSE-UP

UPPER JAW
It may look like a spear, but this pointy bill is the upper jaw of the sailfish! There are tiny teeth on it.

Sailfish swim mostly near the surface of the open ocean and can dive for food as deep as 1,150 feet (350.5 m)!

FISH FAST AND SLOW

Now you know fish can move in many ways. They drift. They dive. They leap. Some are slow, but many are not. The animals in this book are only a handful of the fish that live in fresh water and salt water on Earth. There are more than 34,000 types of fish. Make it your mission to learn even more about how these amazing animals move and how fast they can go!

GLOSSARY

backbone (BAK-bohn) a set of connected bones that runs down the middle of the back; also called the spine

breach (breech) to break through something, like water

camouflage (KAM-uh-flahzh) a disguise or natural coloring that allows animals to hide, making them look like their surroundings

carnivore (KAHR-nuh-vor) an animal that eats meat

cold-blooded (KOHLD bluhd-id) having a body temperature that changes according to the temperature of the surroundings, like reptiles or fish

crustacean (kruh-STAY-shuhn) a sea creature that has an outer skeleton, such as a crab, lobster, or shrimp

dorsal fin (DOR-suhl fin) a single fin on the back of a fish

fry (frye) a very young fish

gills (gilz) the pair of organs near an animal's mouth through which it breathes by extracting oxygen from the water

herd (hurd) a group of seahorses

ichthyologist (ik-thee-AH-luh-jist) a scientist who studies fish

pectoral fins (PEK-tuh-ruhl finz) a pair of fins on either side of a fish that helps it steer

predator (PRED-uh-tur) an animal that lives by hunting other animals for food

prey (pray) an animal that is hunted by another animal for food

scales (skaylz) thin, flat, overlapping pieces of hard skin that cover the body of a fish, snake, or other reptile

shoal (shole) a large number of fish swimming together

wingspan (WING-span) the distance between one end of a wing and the end of the other

INDEX

Page numbers in **bold** indicate images.

A
anglerfish, **4**, 10–11, **10–11**, **30**

B
barracuda, **5**, 24–25, **24–25**

D
dwarf seahorse, **5**, 6–9, **6–9**, **30**
  diet and eating, 7, 9
  movement and speed, 6
  offspring, 8–9
  physical traits, 7–9

F
fish
  common traits, 4
  number of, 30
  ways of moving, 5, 30
flounder, 12–13, **12–13**

G
great white shark, **4**, 22–23, **22–23**

M
manta ray, **5**, 20–21, **20–21**, **30**
moray eel, **4**, 16–17, **16–17**, **30**

S
sailfish, 26–29, **26–29**, **30**
  hunting habits, 28–29
  lifespan, 27
  movement and speed, 26–27
  physical traits, 28–29
sunfish, 14–15, **14–15**

W
whale shark, **4**, 18–19, **18–19**

ABOUT THE AUTHOR

Eric Geron is the author of more than a dozen books. He lives and works in New York City. In the water, he would most likely be able to keep up with an ocean sunfish as it swims along.